A History of the Telephone

From Lovers' Phone to Cell Phone

History of Things
Volume 6

Paul R. Wonning

A History of the Telephone

Published By Paul R. Wonning

Print Edition

mossyfeetbooks@gmail.com

If you would like email notification of when new Mossy Feet books become available email the author for inclusion in the subscription list.

Mossy Feet Books
www.mossyfeetbooks.com

Indiana Places

http://indianaplaces.blogspot.com/

Description

The telephone has come a long way from the primitive "Lover's Phones" invented in 1667 to today's sophisticated cell phone and satellite communication systems.

Also In This Series

A Short History of Transportation

A Short History of Kites

A Short History of Coins

Short History of Candle Making

A Short History of the Game Of Chess

History of the Telephone

Table of Contents

Colonial American History Stories - 1215 - 1664
June 15, 1215 - King John I signs Magna Carta at Runnymede England

A History of the Telephone

Paul R. Wonning

Telephone Etymology

The word "telephone" derives from two Greek words.

French composer Jean-François Sudré first devised the word "telephone," when he developed his "system for conveying words over distance by musical notes." Readers will learn more about Jean-François Sudré in a later article. He used the Greek word "tele", which means far off, afar, at or to a distance and the word "phone," which means "sound" or "to speak."

Lover's Phone

British scientist Robert Hooke's experiments in acoustics led to his developing a device that became known as the "Lover's Phone," or tin can phone in 1667.

How It Works

Hooke used two tin cans with a wire fastened through a hole in the bottoms. If stretched taut, users can talk into one of the cans, allowing another user on the opposite end to place his ear inside the other can and hear what the speaker says. The can acts as a diaphragm which collects the sound wave created by the speaker, converts them mechanical vibrations. The vibrations vary in intensity in response to the speakers words. These vibrations travel along the wire and cause the can to vibrate and covert them back into sound waves, which the listener can hear. The device, known technically as a "mechanical acoustic device," allows people to communicate over longer distances than they could conveniently converse. By tying additional wires, or strings, perpendicular to the main string, other users can join in the network.

Robert Hooke (July 30 - March 14 1703)

The son of John Hooke and Cecily Gyles, Robert was native to the Isle of Wight, which is just off the southern coast of England. His father was a Church of England priest and the head of a local school. He ensured his son's education at the school. Hooke showed interest in observation, mechanical works, and drawing at a young age. He gained admission to Oxford University and later employment to two of Britain's eminent scientists, Robert Boyle and Dr Thomas Willis. Hooke provided valuable assistance to both scientists. He became involved with the Royal Society of London as curator a short time after the Society formed in 1660. Boyle consulted with Thomas Newcomen as he developed his steam engine around 1712. Robert Hooke authored several scientific works during his lifetime and made valuable contributions to mechanics, gravitation, horology (timekeeping), astronomy and paleontology. After his death, he was interred at St Helen's Bishopsgate in London. The location of his grave is uncertain.

"Solrésol"

French composer Jean-François Sudré devised a "telephonic system" by designing a system of musical notes that he used as a code to transmit messages over a distance. He called his system "Solrésol," after the series of notes he used to design his system. The seven notes of the musical scale became words in the system, do (doh), and (re), or (mi), at or to (fa), if (sol), the (la) and yes (ti). His initial thought was that the system would find use on the battlefield, as a bugler could transmit messages to troops in the field. Sudré continued making the system more complex, adding words and nuances as he went. The system eventually failed because of the numerous problems with it. Wind or battlefield noise could interfere with the sound, a listener needed to have musical training to understand the transmission and the size of the instrument limited its range. Sudré spent years, along with two other musicians who accompanied him on tour, trying to "sell," his system. He went as far as designing a gigantic musical instrument, which he called a "telephone", to transmit the messages. The plans and the system died with him.

Jean-François Sudré (August 15, 1787 - October 3, 1862)

Native to Albi, France, Jean-François Sudré studied music as a child and showed enough promise to gain admittance to Conservatoire de Paris on May 12, 1806. At the Conservatoire Sudré studied under such accomplished musicians like François Habeneck and Charles Simon Catel. He migrated to Sorèze and then Toulouse, France in 1818. In Toulouse he opened a music school. By 1822 he returned to Paris to open a music store in which he mostly sold his own compositions. In 1827, Sudré began developing his "telephonic system," which he spent many years developing and promoting. He began developing the Do Re Mi method of notating music sometime in 1829. Sudré spent over three decades trying to have his system implemented. The Paris Exposition awarded him a special prize of 10,000 Francs in 1855 and a jury at the London Exhibition awarded him a Medal of Honor in 1862 for his efforts. After his death, his widow, Josephine, continued his quest by publishing a dictionary of French language Solrésol, the *Langue Universelle Musicale,* in 1866. In spite of this effort, Sudré's musical language has been mostly lost to history.

Semaphore Systems

Though others like Robert Hooke and Sir Richard Lovell Edgeworth had proposed, or used, optical telegraphy before, it was French engineer Claude Chappe and his brothers that developed the first practical system of this type. Beginning in 1790, the Chappe brothers began developing the system that would spread across the world as the first practical long range communication system before the telegraph. Claude Chappe called his new system the Semaphore System, coining the word semaphore from two Greek words, "sêma", which means "sign" and "phorós," which means "carrying."

Military Use

Chappe developed his system to help the French government send long distance messages to military commanders while the allied countries of Britain, the Netherlands, Prussia, Austria, and Spain who sought to squash the French Revolution and restore the monarchy. Lack of reliable communication kept the allied armies from conquering France. Chappe invented his system to help coordinate French forces in the field in their effort to defend France from invasion. Chappe sent his first message on March 2, 1791.

Improving the System

Chappe and his brothers continued improving the system, which evolved to use movable flags mounted on poles. Messages were conveyed by moving the arms to different positions, allowing senders to transmit a coded message to receivers. Each position of the flags indicated a letter of the alphabet or a number. The flags were mounted on towers placed from five to ten miles apart. Observers used a telescope to read the message, and then would relay it on to the next station.

Growth of the System

The French government authorized Chappe to install the first line from Paris to Lille, France in 1792. This line, about 143 miles long, consisted of fifteen stations. The first letter of a message would appear in Paris only nine minutes after transmission began in Lille. It would take about 32 minutes to send a complete message of 36 characters. The French developed several more lines and the system spread to other countries. Napoleon Bonaparte used the system during some of his military campaigns. The first line appeared in the United States around 1801, devised by Jonathon Grout. The line connected Boston and Martha's Vineyard and was used to transmit shipping information. These systems were so successful that Samuel Morse had problems selling his telegraph system, mainly due to concerns about the ease of cutting the telegraph lines and disrupting transmission.

Claude Chappe (December 25, 1763 - January 23, 1805)

The son of Ignace Chappe d'Auteroche and Marie-Renée de Vernay de Vert, Claude was native to Brûlon, Sarthe, France. He was the nephew of famed astronomer Jean-Baptiste Chappe d'Auteroche. One of the first books he read was his uncle's book "Voyage en Siberie," which was his uncle's journal of his trip to veiw the Transit of Venus, awakened his interest in science. His uncle also may have imparted Chappe's knowledge of telescopes that played a role in his development of the semaphore system. Claude's brother Ignace Chappe was a member of the Legislative Assembly during the years of the French Revolution. Ignace played a role by helping get his brother's semaphore system approved.

Early Telegraph Systems

Charles Morrison first suggested what many think was the first concept of an electric telegraph system in 1753

Charles Morrison

Morrison practiced medicine in Greenock, Scotland, appeared in an article of the Scots Magazine on February 17, 1753. In the article, Dr. Morrison suggested using insulated electric wires over a long distance. His system used 26 wires, one for each letter of the alphabet. The wires would have paper indicators on the end with the letter it represented marked. The person on the receiving end could write down the letters as they appeared, deciphering the message. The operator would charge one wire at a time, corresponding to the letter of the word he wanted to transmit. In this way, he believed that messages could be sent over great distances. He also apparently suggested using bells with different tones to send messages. This is the first recorded suggestion of communication using electric wires known to historians.

The First Telegraph

Francisco Salva apparently built a telegraph based on Charles Morrison's concept in 1798 in response to a request from the Spanish king. The king asked him to demonstrate his theory, first proposed in 1795.

Francisco Salva (July 12, 1751 – February 13, 1828)

The son of Dr. Jerome Salvà Pontich and Eulalia Campillo, Francisco was native to Barcelona, Catalonia, Spain. Both his parents were involved in medicine, his mother's family was in the pharmacy business and his father a Staff Physician at Barcelona General Hospital.

Career in Medicine

Salva became interested in medicine at an early age and showed great prowess in the field. At the encouragement of Bishop of Barcelona, Josep Climent, Salva attended the University of Valencia, completing a four year course in three years. He went on to earn a doctorate at the University of Toulouse in France. Salva went on to found a medical school, the Academy of Medical Practice, in 1773. He won several awards from the Paris Society of Medicine for his many accomplishments in the field of medicine.

Telegraph

Salva's interests included far more than medicine. He became interested in the electromagnetic activity and studied the phenomena extensively. On December 16, 1795 he presented a paper entitled "On the application of electricity to telegraphy," to the Academy of Sciences in Barcelona. Charles IV, King of Spain, learned of the paper and invited Salva to demonstrate his device. Apparently, he built a working model that he demonstrated at the Spanish Court. The Madrid Gazette reported on November 25, 1796 that Salva had demonstrated his telegraph system. After several successful runs, system was installed in 1798 that stretched twenty-six miles from Aranjuez to Madrid, Spain, which he based upon the earlier system proposed by Charles Morrison. The limiting factor in Salva's design was the size of batteries available. Salva used a battery based on a design devised by Alessandro Volta, which was one of the earliest versions of a practical battery.

Samuel von Sommering's Telegraph

Using Salva's design Samuel von Sommering constructed a device using improved technology in 1809.

Samuel von Sommering (January 28, 1755 – March 2, 1830)

The son of Johann Thomas Soemmerring and Regina Geret, Samuel was native to Torun, Poland. After completing his primary and secondary education in Torun, Soemmerring studied medicine University of Göttingen. After obtaining his doctorate, Soemmerring became a lecturer, professor and later on a dean of the medical faculty at the University of Mainz. He led an accomplished medical career, becoming the first to discover the macula in the retina of the human eye. He also studied the human brain and nervous system and drew the first accurate drawing of the female skeleton. He also contributed in studies of the lung and human embryo studies. He also worked in the chemistry, astronomy, philosophy and paleontology fields.

Inventor

In addition to his extensive medical contributions, Sommering proved to be something of an inventor. He designed an astronomical telescope and made improvements in the wine industry. In 1809, Sommering built a telegraph, using Salva's design as a guide. He used 35 wires, one for each alphabet letter and number, and a stronger battery than was available to Salva. His demonstration line was about 2.17 miles long at the Munich Academy of Science. The German Museum of Science in Munich has the model on display.

Morse's Electrical Telegraph

May 24, 1844 - Annie Ellsworth of Lafayette Picks First Telegraph Message

Overjoyed that Congress approved the funds to test his telegraphic device, Samuel Morse vowed that Annie Ellsworth, the bearer of his good news, would send the first message by telegraph.

Samuel Morse (April 27, 1791 - April 2, 1872)

The son of Jedidiah Morse and Elizabeth Ann Finley Breese, Samuel was native to Charlestown, Massachusetts. Morse attended Phillips Academy in Andover, Massachusetts and then enrolled in Yale College, from which he graduated with Phi Beta Kappa honors. While at Yale, he had supported himself by painting portraits. He had also attended lectures on the field of electricity, which had fascinated him.

Artist

Artist Washington Allston took note of Morse' work and encouraged him to travel to Britain, where he studied art with Benjamin West. While there, he attended the Royal Academy. Morse developed a distinctive, bold style of portraiture that came into demand when he returned to the United States in 1815. He would paint many portraits of historical figures and historical scenes during his career.

Death of his Wife

Morse married Lucretia Pickering Walker in 1818, with whom he would have one child. Lucretia would die shortly after childbirth in 1825 while Morse was in Washington DC painting a portrait of the Marquis De Lafayette. While working on the painting, Morse's wife gave birth. He received a letter from his father informing him that Lucretia was recuperating from the birth. The next day, he received another letter from his father informing him that Lucretia had died. Morse abandoned the painting and traveled home by train, arriving after his wife was already buried. Overwrought that he had not found out that his wife was deathly ill until she was already dead and buried, an anguished Samuel Morse abandoned a painting career and developed a means of instant electronic communication.

The Telegraph

Morse received inspiration from conversations with passengers on a boat on which he traveled on a return trip to the United States from a trip to Europe in 1832. Michael Faraday, a pioneer in electrical science, had just invented the electromagnet. Through conversation, Morse learned how electricity worked. He wondered if it would be possible to send a coded message along a copper wire using the electromagnetic device. Morse, in consultation with a colleague Professor Leonard D. Gale, began developing his telegraphic system. By 1837, he felt confident enough to apply to the Federal Government for an appropriation to set up a test line. However, his application arrived at the same time as the Financial Panic of 1837 hit and the Government turned him down. He continued to work on, and improve his system before applying again in 1843.

Success

After demonstrating the device in Congress, many key Congressmen were still reluctant to approve the $30,000 Morse required to set up an experimental thirty-eight mile line from Baltimore to Washington, DC. During negotiations over the bill, a frustrated Morse returned to his hotel in Washington, convinced that Congress would reject his application. He began preparations to return to his home in New York. His old college friend, former Commissioner of Patents Henry L. Ellsworth, continued to work on the bill. In the wee hours of the night, the bill did pass. Ellsworth sent his sixteen-year-old daughter, Annie, to Morse's hotel room to deliver the good news. Upon hearing the news, an elated

Morse vowed that Annie would have the honor of composing the first message. He did send the message Annie composed, "*What Hath God Wrought,*" from Washington to Baltimore on May 24, 1844.

Annie Goodrich Ellsworth (1827 - January 21, 1900)

The daughter of Henry L. Ellsworth and Nancy Allen (Goodrich) Ellsworth, Annie was native to Hartford, Connecticut. Henry Ellsworth moved his family to Lafayette, Indiana after finishing his term at the Patent Office in 1845. After the move, Annie met and married Roswell Chamberlain Smith. Annie would have six children. Her son, Van, founded Roswell, New Mexico in 1871, naming it for his father. She would reside in Lafayette until 1870, when she moved to New York.

Innocenzo Manzetti's "Speaking Telegraph"

An Italian inventor laid the foundation for the telephone with his speaking automaton in 1849.

Innocenzo Manzetti (March 17, 1826 - March 15, 1877)

Native to the Italian town of Aosta, which is near the Italian/Switzerland border, Manzetti attended the primary schools in his hometown. After his primary education, he enrolled in the He displayed an early interest in mechanics and attended Saint Bénin Boarding School, a Jesuit boarding school, and then furthered his education in Turin, Italy. He returned to Aosta after receiving a diploma as a land surveyor.

Automaton Flute Machine

Manzetti first broached the idea of a "speaking telegraph" in 1843, however he did not pursue his idea until he wanted to give a voice to his first notable invention, a clockwork-driven automaton that could play the flute in he presented to the public 1849. The “flute player” was a man-sized machine that sat on a chair. He covered the machine with suede and even gave it porcelain eyes for realism. The automaton could move its arms and legs using various levers, connected rods and compressed air tubes. It could roll its head and roll its eyes. The machine used a pre-recorded cylinder to play up to twelve different tunes while its fingers moved. It could even take off its hat and bow its head after completing a performance. Manzetti used a spring-like device to wind it up like a clock to supply power. He exhibited his mechanical marvel at the London World’s Fair in 1851.

Talking Machine

Manzetti continued improving his machine over the years and in 1864 reports appeared that he had incorporated a speaking telegraph into the machine to allow it to talk. News reports at the time stated he could transmit his voice to the machine over wires similar to those used by the telegraph and that his device could "talk," using Manzetti's voice. Manzetti did not patent his device and few official records of its performance exist.

Other Inventions

Manzetti devised many other devices in addition to his automaton. These included a wooden flying parrot built as a toy for his daughter, a hydraulic pump to remove water from mines and a set of geodetic instruments that he used in his survey work. He also built a a steam-powered car, a pasta making machine, a filtering system for public drinking water and a pantograph to make reproduced drawings on bas-reliefs on ivory, marble or wood.

Little Recognition

Manzetti received little recognition for his "talking machine," until recent years.

Johann Philipp Reis' Talking Violin

Johann Philipp Reis devised his "telephon"in 1861.

Johann Philipp Reis (January 7, 1834 – January 14, 1874)

The son of Marie Katharine (Glöckner) and Karl Sigismund Reis, Johann was native to Gelnhausen, Germany. His mother died while he was still a baby, thus his father, a baker, sent him to his parents to raise. His elementary education took place in the common schools of Gelnhausen. He later attended the Garnier's Institute, in Friedrichsdorf, after his father died in 1844. He later attended the Hassel Institute near Frankfurt. After finishing a year of military service, he took employment as a teacher at the Garnier's Institute.

The Reis Telephone

Reis developed a theory that electricity could pass through space similar to the way sound moves. He began experimenting with a device he called a "*telephon*" and demonstrated his model to the Physical Society of Frankfurt in 1861. He used a wooden, human ear shaped device as the microphone and a sausage skin as the diaphragm. He fastened a platinum strip to the diaphragm for one contact and used an adjustable bead as the other contact. This apparatus was contained in a closed wooden box at the back of the ear. He mounted the contacts to a copper wire, which ran a short distance to a battery pack. From there, the wire was wrapped around a knitting needle, which was encased in silk. This was placed inside a violin. When someone spoke into the ear, the sound waves caused the diaphragm to vibrate, causing the contacts to fluctuate. This sent a current down the wire, powered by the battery, to the knitting needles, which vibrated in response to the fluctuating current. The violin acted as a resonance chamber, allowing people to hear the sound. The device worked well enough, but speech was hard to understand. He had based his design on Charles Bourseul's ideas. Alexander Graham Bell later admitted that he used Reis ' design as inspiration for his device. Reis' telephon was regarded as little more than a toy, but it did prove as a forerunner to Bell's telephone.

First Microphone

Charles Bourseul constructs what many believe was the first microphone in 1854

Charles Bourseul (April 28, 1829 – November 23, 1912)

The son of a French army officer, Charles was native to Brussels, Belgium. The family moved to Douai, France while Charles was young. He gained employment in the French telegraph system. He began making improvements to the system. His work with the telegraph led to his speculation that if he could construct a device that had a flexible disc would vibrate in response to the human voice. If he connected the disc to a sensitive electrical switch, the connection would break and reconnect in response to the sound waves, sending a message along the wire. He published an article in the magazine "*L'Illustration de Paris,*" in 1854 that outlined his theory. Apparently, he built a prototype that worked, but he could not get the receiver to function properly.

Antonio Meucc The First Telephone?

History has credited Alexander Graham Bell with inventing the first telephone, however historical evidence indicates that Antonio Meucc's invention preceded Bell's device.

Antonio Meucc (April 13, 1808 - October 18, 1889)

The son of Amatis Meucci and Domenica Pepi, Antonio was native to Florence, Italy. Meucc attended the Florence Academy of Fine Arts where he studied design and engineering. After completing his studies, he gained employment at various theaters in Italy. In 1834, Meucc developed an acoustic pipe telephone that allowed stage workers to communicate with people in the control room. Political turmoil in Italy caused him to immigrate in 1835, when he migrated to Havana Cuba. In Havana, Meucc worked as a scenic designer and stage technician. Meucc began his inventing career in Havana while he worked in this capacity.

Accidental Discovery

Meucc developed a new metal galvanizing method for the Cuban military and then began investigating the use of electric shocks to treat pain. While using one of the devices to treat a patient, Meucc accidentally discovered the voice carrying properties of copper wire. Meucc wanted to investigate the phenomenon further, so in 1850 he and his wife moved to New York City, taking with them their substantial savings.

Investor and Inventor

Meucc wanted to pursue a career as an inventor, encouraged by the success of Samuel F. B. Morse . Meucc invested his savings in a tallow candle factory, hoping to use the profits to fund his inventing career. He continued studying the electromagnetic effect of voice transmission over copper wires and by 1856, he had installed an electromagnetic telephone that ran from his work shop to his adjoining house.

Bankruptcy and Failure

Meucc wanted to build a larger prototype and file a patent for his device, however the candle factory went bankrupt, leaving him bereft of the funds he needed. For the next several years, he struggled to find investors and to develop his idea further. He finally succeeded in filing a patent caveat (temporary application) on December 28, 1871. He continued working on other inventions. Improper descriptions of his device on the patent applications caused the application to expire without his receiving approval. Controversy continues to swirl among historians over whether Alexander Graham Bell or Antonio Meucc was the first to develop the telephone. Bell receives the credit, because he successfully applied for his patent in 1876, however historical evidence suggests Meucc really invented it first.

Alexander Graham Bell and the First Telephone

Historians generally credit Alexander Granham Bell with the first telephone.

Alexander Graham Bell (March 3, 1847 – August 2, 1922)

The son of Alexander Melville and Eliza Grace Bell, Alexander was native to Edinburgh, Scotland. Bell's parent's home schooled him until he was eleven years old, after which he attended the Royal High School in Edinburgh. At age eleven, he visited a flourmill with his friend, whose father owned the mill. After watching the laborious process of husking chaff from the wheat, Alexander invented a machine that did it faster and with much less effort. His friend's father used the machine in the mill for many years. His father allowed him to set up a workshop after this so he could continue to work on various projects. He received poor grades and had poor attendance at the Royal High School, leading him to abandon formal schooling at fifteen.

Elocution and Communication

Bell's family had a history of mastering the arts of elocution, and Alexander was no different. He enrolled at the Weston House Academy at Elgin, Moray, as a student teacher of elocution and music. Bell had also displayed a talent for music, mastering the piano at a young age. His mother began going deaf while he was young, causing Alexander to learn a finger-tapping language so he could communicate with her. Elocution and the fascination with communication would drive Alexander's quest to invent his telephone device. He spent many hours in his workshop conducting experiments with sound.

Move to Canada and Work in the United States

In 1870, the Bell family moved to Canada, staying with a family friend in Paris, Ontario until they purchased a farm near Tutelo Heights. Alexander set up a workshop and continued working on his various electricity and sound projects. In 1872, he opened the School of Vocal Physiology and Mechanics of Speech in Boston, Massachusetts and later became a Professor of Vocal Physiology and Elocution at the Boston University School of Oratory the next year. He opened a workshop in Boston in 1874 to work more on the harmonic telegraph machine he had been working on for a number of years.

The Telephone

Though he lacked funds, support from colleagues allowed him to hire Thomas Watson as an assistant to work on acoustic telegraphy. An accident in the laboratory allowed Watson to detect a sound traveling over the wires in the workshop, which led Bell to develop his sound-powered telephone. The development led to a "patent office race," between Elisha Gray and Bell. Some controversy still exists over who actually received the first patent, Gray or Bell. Nevertheless, Bell received his patent on March 7, 1876. Bell went on to found the Bell Telephone Company. Bell and his assistants continually worked to improve the device and ten years after he received his patent over 150,000 people in the United States owned telephones. Bell refused to install one in his workshop, considering them a nuisance and an interruption to his work.

Electromagnetic Telephone

The electromagnetic telephone devised by Bell consists of two main parts, the transmitter and the receiver. The transmitter is the ear of the device, with a thin metal disc that is called a diaphragm. When a person speaks into the "ear", the varying intensity of the speaker's voice causes the diaphragm to vibrate at differing levels in response to the sounds. Behind the diaphragm lies a small compartment of carbon grains that respond to the pressure caused by the vibrating diagram. Electric current, supplied by a battery located at the phone company, flows through the carbon grains. The current varies according to the pressure of the carbon grains and flows thought the wire to the receiver of the listener. The receiver consists of a metal disc, also called a diaphragm that is attached to an electromagnet. The electric current coming from the transmitter passes through the electromagnets coil becomes magnetized. The varying electric current causes the electromagnet to vibrate against the diaphragm, recreating the sounds uttered by the speaker. The listener can detect this sound as words by holding the receiver against his ear. Thus, two people miles apart could carry on a conversation. The sounds reproduced by the early phones were so weak that a listener had to hold their ear quite close to the receiver to hear them. However, Bell and other technicians improved the device over time allowing better sound reproduction.

Other Inventions

Bell invented many other things over his lifetime in addition to the telephone. These included a wireless telephone, a phonograph, metal detector and a hydrofoil watercraft. He also contributed to the medical field and aeronautics. After his death on March 3, 1847, every telephone in the United States went silent after his funeral service for one minute to honor his invention.

Elisha Gray's Telephone

Both Gray and Alexander Bell filed for patents on the same day; however, Bell won the court case giving him the privilege of being known as the first.

Elisha Gray

The son of Quaker parents, David Gray and Christiana Gray, Elisha was native to Barnesville, Ohio. Gray apprenticed to a blacksmith before enrolling at Oberlin College on Oberlin, Ohio. Gray attended classes and pursued his interest in electricity. He did not graduate from the college, however he established a laboratory there and taught classes on the subject.

The Telephone

Gray is best known for his invention of the telephone and his failed dispute with Alexander Graham Bell over the patent of the device. Gray worked on the telephone in secret until February 14, 1876, when he filed a patent for it. Alexander Graham Bell filed the same day, sparking a dispute over who filed their patent first. Bell won the court case and received credit for the telephone, despite evidence that Gray may have filed first.

Other Inventions

Gray received patents for over seventy devices during his lifetime. These included pioneer fax machines, a self-adjusting telegraph relay, a microphone printer and the telautograph, which could transmit handwriting through a telegraph system.

In addition to his inventing career, Gray wrote four books:

Experimental Researches in Electro-Harmonic Telegraphy and Telephony, 1867–1876 (Appleton, 1878)

Telegraphy and Telephony (1878)

Electricity and Magnetism (1900) and

*Nature's Miracles (*1900)

1877 - First Long-Distance Telephone Line

About a year after Bell received the patent for his telephone, the Ridge Telephone Company deployed the first long distance telephone line in California. The line stretched fifty-nine miles from French Corral to French Lake in Nevada County on the eastern edge of northeast California. Workers strung the line in 1877 over poles and through trees connecting the Milton Mining and Water Company and other water companies. An historical marker on Pleasant Valley Road near French Corral, California denotes the spot.

Bell Patent Association

Alexander Graham Bell and his assistant Thomas Watson established a verbal trusteeship and a partnership that the partners used to protect any patents they developed. The Bell Patent Association evolved into the Bell Telephone Company in 1877.

Water Microphone

The water telephone preceded the carbon microphone, developed by Thomas Edison on March 10, 1876, and provided the inspiration Edison used for his device. A water telephone uses a rod or needle suspended in weak acid solution, composed of acid and water. The needle is attached to a metal diaphragm, which vibrates in response to sound. The vibrating needle or rod causes fluctuations in the circuit's current, causing variations in the flow of electricity. Elisha Gray first described the water microphone and Bell used a different version, using mercury as the liquid, in his patent application. No one actually used the device, as it was impractical. However, Thomas Edison used the concept to design his carbon microphone in 1886.

Pulsion Telephone

Based upon the technology of the "tin can phone," Lemuel Mellett established the Pulsion Telephone Supply Company in 1888. The system was installed in several railroads. The main advertising points were its independence of electricity and ease of installation. The main components were a copper wire, or twisted steel wire, stretched tight and attached to two discs. The discs had several spiral springs in a case attached to the wire. When someone spoke into one disc, the sound vibrated along the wire and into the opposite disc. The springs magnified the sound, allowing the listener to hear messages transmitted up to three miles from the transmitting disc. A hat laid along the exposed wire at any point allowed anyone to hear the message. The company saw some success, as experiments indicated that the wire could pass through the branches of a tree or underwater with no interference in sound quality. The systems were cheap to install and needed no batteries. Improved telephone technology caused this system and many others like it to go out of business by the turn of the century.

Carbon Microphone

Three men developed the carbon microphone independently, Thomas Alva Edison, David Edward Hughes and Emile Berliner sometime around 1878. Edison was the first to patent the device, so he has received the credit, though evidence suggests that David Edward Hughes actually invented the microphone first.

Thomas Alva Edison (February 11, 1847 – October 18, 1931)

The son of Samuel Ogden Edison Jr. and Nancy Matthews Elliott, Thomas was native to Milan, Ohio. Edison's mother, an accomplished schoolteacher, home schooled him, using the books R.G. Parker's School of Natural Philosophy and The Cooper Union for the Advancement of Science and Art as his textbooks. She helped him develop his desire for self-education, which he used to great advantage later in life. Early in life, Edison developed scarlet fever and several ear infections that led to a loss of hearing. When he was seven years old, the family migrated to Port Huron, Michigan near a train station.

Early Life

At age twelve, Edison noticed the news stories arriving at the train stations by teletype. He compiled these stories into a small newspaper that he published in a periodical he called the *Grand Trunk Herald*. His parents allowed him to sell these newspapers on the trains. His interest in experimenting led him to put a small laboratory in a train baggage car. He accidentally set the car on fire when some chemicals combusted, leading to his expulsion from the train. He continued to sell the newspapers at the train station. During this time, he saved a three-year-old boy from being run over by a train and in response; the grateful father taught the fifteen-year-old Thomas the basics of operating a telegraph. This led to his first job in a telegraph office, eventually gaining employment in the Western Union Company in Boston.

Inventor

Edison began his inventing career at twenty-two years old after he moved to New York. The Gold and Stock Telegraph Company paid him $40,000 for his invention, an improved stock ticker. Edison would go on to patent 1,093 inventions including the light bulb, the carbon microphone, the phonograph and a motion picture camera. At his death at his New Jersey home, scores of communities around the United States and the world dimmed their lights for a few minutes in his honor.

The Carbon Microphone

The carbon microphone has a layer of carbon granules sandwiched between two metal discs. The outer disc acts as a diaphragm, vibrating as a person speaks into it. The carbon granules' resistance varies in response to the sound waves, changing the current. The electrical system inside the phone changes this varying sound signal to an electrical signal. The signal travels through the phone system's copper wire, delivering it to the receiver on another phone. The carbon microphone was instrumental for telephones, radio and as a sound amplifier.

Telephone Exchange

Before the evolution of the telephone switchboard, an operator had to connect them manually to each other for them to work. The most common use was for a business owner to connect the telephones to a phone in their home and one in their business. It was impractical to connect the phones with other phones.

First Telephone Exchange

Hungarian engineer Tivadar Puskás came up with the idea of a telephone exchange that would allow several phones to connect with each other while he worked with Alexander Graham Bell. During an April 27, 1877 lecture about telephones George W. Coy drew inspiration to build the first telephone exchange. Bell demonstrated a three-way connection during his lecture, inspiring Coy to apply for, and receive, a franchise from the Bell Telephone Company to build the exchange on November 3, 1877. With less than forty dollars in materials, which consisted of bolts, teapot lid handles and wire, Coy built the device and began operation on January 15, 1878. Coy's District Telephone Company of New Haven could connect sixty-four customers. The exchange could only handle two simultaneous telephone calls and required six connections to make a call. The company began with twenty-one subscribers who paid $1.50 a month. Businesses, physicians, police department, and the post offices made up the bulk of the listings, with only eleven private individuals subscribing to the service. By 1880, the company received the right from Bell Telephone to connect telephones in Connecticut and western Massachusetts.

Sound-Powered Telephones

Developed in the 1890's, sound powered telephone system use a mechanical transducer, which is a device that converts one form of energy to another, to convert sound waves into an electrical signal. The signal is carried waves over a copper wire into another transducer that that converts the electrical signal back into sound so the listener can hear the words. Sound powered systems need no electrical power source, so they are ideal under many naval, military and emergency vehicle applications. Sound powered systems are sort of like an advanced version of the "tin can phone," in which the transducer replaces the tin can as the diaphragm and the wire replaces the string. They saw their first deployment on United States Naval ship during World War II and are still in use in many places.

Early Ringers

The earliest telephone users alerted other users on the line or the switchboard operator that they wanted to talk by whistling into the transmitter. Two types of ringers evolved over the next few years to replace the primitive whistling system. After exchanges began appearing, those that used a common battery began installing bells, operated over a separate line that the operator used to ring the phone of the second subscriber when the first one wanted to talk. Exchanges that still used the local battery system used a magneto-type switch mechanism. The magneto switch disconnected the talking circuit when the user began turning the crank. This sent an A/C current along the line, alerting the operator that a subscriber wanted to talk. The operator connected the two parties and rang their line. When the other party picked up, the magneto disconnected and the line became active for talking. Some operators developed ringing codes to denote individual subscribers, emergency phone calls or urgent calls. In some areas, this type of phone system survived into the 1950's.

Strowger Switch

An undertaker, Almon Brown, invented the Strowger Switch, which was the first automated telephone switching system.

Almon Brown Strowger (February 11, 1839 – May 26, 1902)

Native to Penfield, New York, near Rochester, historians know little of his early life except that he and his brother would spend time trying to figure out how to build a machine that their mother bestowed upon them rather than doing the task. He spent time as a schoolteacher, enlisted in the Union Army during the Civil War and returned to teaching school after the war. He moved around a lot and ended up working as an undertaker. Again, his location at the time he conceived the idea is uncertain, however he listed his residence as being Kansas City, Missouri.

Diverted Business

Historical lore suggests that the idea for the automatic switching machine came to him one morning as he read the newspaper and noticed that a friend had died and another local undertaker was taking care of the funeral arrangements. With a little detective work, he found that that undertaker's wife worked in the telephone exchange. He suspected that she was diverting business to her funeral parlor.

Inventing the Automatic Switch

At first he became enraged by the unethical behavior of the operator, however her behavior soon set him off on another course. He decided he would come up with an idea that would eliminate the need for a telephone operator. He drew his idea from paper shirt collars he used in his business and some straight pins. He stuck the straight pins in the collar with the sharp end pointing toward the center. Examining him, he mused that an electrician should be able to rig an electromagnetic device that would connect the desired pins together with the proper signal. He enlisted his brother, Arnold and his nephew, William, to help him build the device. Both men had some knowledge of electricity and with some help from others in the field, soon invented the device. The men formed the Strowger Automatic Telephone Exchange Company and filed there for their patent, which they received on March 10, 1891.

Experimental Installation

Strowger had moved to Laporte, Indiana in the meantime. He installed an experimental version of the device in the telephone exchange in Laporte in 1892. Since it was experimental, Stowger and his company did not charge subscription charge. The Bell Company found out he was operating outside their system and filed a lawsuit, which they dismissed when Stowger informed them that it was experimental and they were offering the service free. After the system proved satisfactory, Stowger sold it to some Laporte businessmen who did sell subscriptions, apparently with the blessings of the Bell Company.

Rotary Dial

Almon Brown Strowger patented the rotary dial on November 29, 1892 as an essential part of his automated telephone switching system. The rotary dial functions by sending a specific electronic pulse in response to the position the dialer stops to the telephone exchange. Each spot on the dial has a number assigned to it. When the dialer stops at the number and releases the dial, it returns to the home position, sending this signal to the exchange, which interprets the signal as a number. By sending a specific set of numbers, the rotary dialer tells the automatic exchange to what number to connect the dialer. The first rotary dials did not have the familiar finger hole in the dial, nor did they have a spring return. The design made it hard for users to dial the numbers. The rotary dial evolved to include the spring and finger holes by 1904, however the dial system did not see widespread installation until the 1930's.

Candlestick Telephone

The candlestick telephone became common during the 1890's and early Twentieth Centuries. The telephone's name derived from its resemblance to a candlestick. The microphone was at the top of the "candlestick," and the receiver suspended on a hook on the side. When the user placed the receiver in the hook, it depressed a switch that disconnected the phone. The candlestick required a separate battery box to power it and a separate ringer, as neither was on the phone. After the rotary dial system appeared, candlestick phones included a rotary dial on the base of the telephone. Many manufacturers made these phones, including Western Electric, Kellogg Switchboard & Supply Company, and Stromberg-Carlson, thus there were several different variations of these phones.

First Transcontinental Telephone Call

The first "official," transcontinental telephone call took place between Alexander Graham Bell and his assistant, Thomas Watson, on January 25, 1915. The first call had actually occurred in July 1915 as a test call between Theodore Vail, president of AT & T and an assistant on the west coast. Alexander Graham Bell picked up a telephone in New York on January 25, 1915 and completed the first official call. The company had picked the date to coincide with the Panama-Pacific International Exposition held in San Francisco that year to celebrate the opening of the Panama Canal. During the conversation, Bell transmitted his famous question, "Mr.

Watson, come here. I want you," posed so many years before, to his assistant, over 2500 miles of telephone line to Watson in San Francisco. The symbolic call launched the age of instant, coast to coast, communication.

Radiotelephone

In 1927, radiotelephone service was established, linking America with Great Britain. Atmospheric conditions influenced service, making calls subject to fading out and static. About 3000 calls were made the first year of the service. Radiotelephone systems still exist in rural areas and other places not accessible to normal cellular phone service.

Bell's Model 102 Telephone

The Model 102 Telephone began superseding the candlestick telephone sometime around 1927. The Model 102 Telephone initiated the beginning of the telephone design that would dominate the telephone industry for several decades. The telephone featured a handset that included both the receiver and transmitter. Thus, the user could pick up the handset and be able to talk and listen with one device. The dial was on the base and the ringer inside the base, eliminating the need for a separate ringer box. Common battery power was becoming more common, so the need for a separate battery box was also becoming obsolete.

1955 - Transatlantic Telephone Cables

The United States Post Office, the Long Lines Department of American Telegraph and Telephone Company, Bell Telephone Laboratories and the Canadian Overseas Telecommunications Corporation signed an agreement on December 1, 1953 that committed them to laying a communications cable that would connect the United States, Canada and England. The idea had been floated as early as the 1920's; however, the technology did not yet exist to make the concept feasible. By the mid 1950's cable and electronic capabilities had advanced to a point that it was possible to do. Work began with specialized ships laying the cable beneath two miles of ocean in February 1955. The first call was made during an inaugural ceremony on September 25, 1956. The 1500 nautical mile cable provided over thirty telephone circuits. During the first day, almost 600 calls were completed on the new cable.

Push Button Telephone

Western Electric began experimenting with push button telephones in 1941. The Bell System began experimenting with customers around 1960 and introduced the first push button telephone, called the Touch-Tone, to consumers on November 18, 1963. The Touch-Tone took several decades to supersede the rotary dial, however it was not until the 1990's that push button telephones formed the overwhelming majority of telephones in use. The first push buttons only had ten buttons, one for each digit. Later phones had twelve with the addition of the * and # signs.

Party Lines

Up until the 1980's, many telephone subscribers had party lines. Although private line service was available for an extra fee since the 1880's, many subscribers elected to have the cheaper party line service. In some rural areas, party lines were the only telephone service available. On a party line, several subscribers connected in a network. Each person on the line could hear any other subscriber's telephone conversation by simply picking up the receiver and listening. Party lines served as a means of passing news and gossip around the neighborhood. Since there was no privacy on a party line, conversers had to watch what they said, as there was always the chance the conversation could be overheard. By the 1980's, new telephone switchboards could no longer accommodate the party line. Accessories like answering machines and computer modems were also not compatible with party lines. Telephone companies gradually phased the party line out.

Cellular Phone History

Cell phones are small devices using radio waves to connect with a cell tower, which then connects the caller with the receiver. An understanding of the history of the mobile telephone involves a basic understanding the history of radio.

Electromagnetic Wave Transmission

Electromagnetic Waves are synchronized oscillations of electric and magnetic fields that move at the speed of light (186,000 miles per second) through a vacuum. The waves create what scientists call electromagnetic radiation. Radio waves are a form of electromagnetic radiation that can move through the atmosphere. A scientist named Henrich Hertz did the first important work with this medium in 1885. A man named Nikola Tesla also performed work in wireless transmission.

Henrich Hertz (February 22, 1857 - January 1, 1894)

The son of Gustav Ferdinand and Anna Elisabeth Pfefferkorn Hertz, Henrich was native to Hamburg, Germany.

Education

Hertz attended grammar school at the Dr. Wichard Lange School in Hamburg, where he became a top student. His mother arranged for him to take drafting lessons, as he had displayed a talent for drawing and making things. He left formal education at age 15 for home schooling. A tutor taught him Greek and Latin, which he would need to be admitted to a university. He also received lessons from a math tutor. He began university studies in 1874, attending engineering studies in several German cities. In 1880, he earned his doctorate in engineering from the University of Berlin after submitting a masterful thesis on electromagnetic induction.

Marriage and Career

That same year he married Elisabeth Doll, with whom he would have two children. Hertz attained a professorship at the University of Karlsruhe. At the university, Hertz performed experiments in the field of electromagnetic wave transmission, building a machine that produced and detected radio waves in 1886. Even though Hertz could see no practical use for his discovery, the International Electro technical Commission used his name to measure the frequency of radio waves, the Hertz (Hz) unit. In addition to his important work in the field of electromagnetic transmission, he had major contributions to the field of meteorology, contact mechanics and cathode ray tube, an important component in early televisions.

Nikola Tesla (July 10, 1856 - January 7, 1943)

The son of Milutin and Đuka Mandic Tesla, Nikola was native to Smiljan, Serbia. Tesla attended primary school in Smiljan and then nearby Gospic when the family moved there. He enrolled in high school at the Higher Real Gymnasium in Karlovac, Croatia. He graduated in 1873 after demonstrating extraordinary mathematical skills.

Illness

After returning home, Tesla contracted cholera, a disease that almost killed him. He was laid up for about nine months. After his recovery, he enrolled at the Austrian Polytechnic in Graz, Austria. During this time at the institute, he acquired a gambling habit and lost his allowance and tuition money. He left school in December 1878, severing all ties with his family to prevent their learning of his situation.

Early Life

By 1882, two uncles had rescued him and he had worked at the Budapest Telephone Exchange, quitting that to take a job at the newly established Continental Edison Company. During his time at Edison, he began work on an arc lighting system and moved to the United States in 1884. He quit Edison to form his own company in 1885. Over the next few years, he did serious work on an A/C induction motor and a steam-powered oscillating generator. During this time, he also developed his Tesla coil. In 1895, Tesla began experimenting with X-Rays. In 1898, he built a remote controlled boat, an idea he tried to sell to the military, without success.

Wireless Power

From the 1890's until 1906, Tesla began experimenting with the concept of wireless transmission of electrical power. His idea was to use large coils of copper wire to transmit electrical power wirelessly from a transmitter to a receiver. He envisioned using the atmosphere to conduct these signals, amplifying them with a series of airborne balloons high in the atmosphere. He managed to obtain funds from millionaire J. Pierpont Morgan and built the Wardenclyffe Tower in Shoreham, New York. Guglielmo Marconi had already begun his radio experiments and was on the verge of applying for a patent. Tesla suspected that Marconi had copied his designs and strove to become the first to transmit radio signals. Marconi beat him, however, conducting a successful test in December 1901. The experiments with wireless power failed to attract investors and Tesla, mired in debt, spent the remainder of his years entangled in lawsuits with Marconi and others. He passed away on January 7, 1943.

Guglielmo Marconi and the First "Wireless Telegraphy" Device

Guglielmo Marconi sends the first successful radio signal in 1895.

Guglielmo Marconi (April 25, 1874 - July 20, 1937)

The son of Giuseppe and Annie Jameson Marconi, Guglielmo was native to Bologna, Italy. Marconi apparently was a poor student in school. During his early childhood, he lived in England with his mother, however he returned to Italy when he was six. Since the boy showed an intense interest in electricity, his neighbor in Italy, physicist Augusto Righi allowed him to listen to lectures at the University of Bologna. Many of these lectures detailed the work of Heinrich Hertz. Marconi also attended lectures at other universities in Italy as well.

Radio

During the early to mid 1890's when he was in his late teens, Marconi began working on "wireless telegraphy." In 1894, he built a "storm detector" which detected the radio waves sent out by lightning and set off an alarm. Marconi continued working in his workshop in Villa Griffone in Pontecchio, Italy. He built most of his own equipment, aided by his butler Antonio Mignani. In 1895, he successfully sent a radio signal over a distance of a mile and a half. He traveled to England in 1896, where he received the first patent for a piece of radio equipment.

Wireless Telegraph and Signal Company

After receiving the patent, Marconi set up his first company, the Wireless Telegraph and Signal Company in Chelmsford, England. Marconi kept working to improve his device and on December 17, 1902 sent the first radio transmission across the Atlantic Ocean. On January 18, 1903, he sent a radio message from President Theodore Roosevelt in Washington DC to King Edward VII in England. Reliable radio transmission was still not reliable; however, Marconi kept working on the new medium and by 1907 had established subscription news services for ships at sea.

Marriage

Marconi married Beatrice O'Brien in March 16, 1905. The couple would have four children, three daughters and a son. The couple would later divorce, and since Marconi was Catholic, had the marriage annulled so he could marry Maria Cristina Bezzi-Scaliin 1927. They would have one daughter.

Wireless at Sea

Marconi's invention proved its value when the *Titanic* sank in 1912 and the *Lusitania* in 1917. The loss of life on both ships would have been much higher if radio operators had not used the radios in the ships to alert rescuers of their plights. Marconi was in charge of the radio service for the Italian military during World War I, his new technology playing a role in the war. Marconi died of a heart attack in Rome on July 20, 1937. The British Broadcasting Network honored him with two minutes of radio silence during his funeral. Marconi is interred at the Villa Griffone at Sasso Marconi, Emilia-Romagna.

Audion Vacuum Tube

Developed by inventor Lee de Forest in 1906, the Audion Vacuum Tube was the first electronic device that could amplify a signal. It consisted of three elements, a metal plate, a control grid and a carbon filament enclosed in a gas filled glass tube.

Control Grid

Usually in the form of a wire helix shaped spiral or screen, the control grid can be made from either metal or gauze. If it is metal, the components are usually nickel, molybdenum or an alloy of these. In many of the tubes, the molybdenum has gold plating. The metals used in the grid must be able to withstand the high temperatures generated within the tube. In a typical tube, the grid is positioned between the anode (plate) and cathode (filament).

Anode (Plate)

Usually constructed from molybdenum, the plate acts as a heat sink that dissipates heat generated in the vacuum tube. Sometimes the plate is flat in shape, but may be cylindrical or box shaped.

Cathode (Filament)

The first filaments were made from carbon; later ones were made from tungsten. The filaments construction is similar to the filament used in an incandescent light bulb, from which it evolved.

The Gas Envelope

The covering of a tube is glass that is sealed at the base. Before sealing, a pump removes the air from the envelope and replaces it with a gas. The gas functions to prevent oxygen from oxidizing or evaporating the filament. The gas used can be argon, nitrogen, krypton or other inert gas.

Basic Operation

At a very basic level, the cathode emits electrons and heat when electric is applied to it. The electrons strike the grid, which controls the electron current. The anode plate then absorbs this power. The tube de Forest devised could detect radio code and voice messages. The Audion Vacuum Tube evolved to form the foundation of all later electronic devices.

Lee DeForest (August 26, 1873 - June 30, 1961)

The son of Henry Swift and Anna Margaret Robbins Deforest, Lee was native to Council Bluffs, Iowa. DeForest spent two years studying at the Mount Hermon Boys' School in Mount Hermon, Massachusetts, after which he enrolled at Yale University's Sheffield Scientific School in New Haven, Connecticut in 1891. DeForest's experiments in electricity sometimes blew fuses in the building, a result that eventually got him expelled.

Military and Early Career

He entered the army at the beginning of the Spanish American War, but the war ended before he saw any action. DeForest, interested in the new wireless telegraphy field, DeForest attempted to find work with both Nikola Tesla and Nikola Tesla, but was unsuccessful. DeForest worked in the field at various companies over the next few years, including the Western Electric Company, American Wireless Telegraph Company and the Armour Institute.

The Audion Tube

In 1902, he teamed up with Abraham White, who helped him set up the American DeForest Wireless Telegraph Company. It was during this time that he developed the Audion Vacuum Tube in 1906. DeForest and the main stockholders in the company came into conflict, so DeForest sold his stock and resigned from the company. He established the Radio Telephone Company to promote his various inventions. His inventions included a system that allowed sound to be recorded on film, opening the door for "talking" motion pictures between 1923 and 1927. He also developed several medical devices and did military research during World War II. His work on the Audion Vacuum Tube has led many to call him the "Father of Radio."

Radio Act of 1912

Approved by Congress on August 13, 1912

The sinking of the Titanic began on April 15, 1912 launched an investigation into the tragedy. The investigation revealed, among other things, that if the Titanic had not had radio equipment installed, there would have been no survivors. The 700 people that lived owed their lives to Guglielmo Marconi's device.

Radio Act of 1912

Congressional legislators initiated the bill, which passed Congress on August 13, 1912, in the wake of the catastrophe. The Act stipulated:

All ships at sea with radio equipment have trained radio operators

Ships at sea with radios must monitor emergency radio frequencies

All radio stations in the United States must have a license

The law set radio frequencies these stations must use

The law also defined various wavelengths and set general restrictions on private radio stations. The U.S. Department of Commerce and Labor were authorized to issue the licenses and impose fines and penalties on those that did not comply.

Radio in the First World War

The radios that served the land armies in World War I were still primitive units. Limited in range to about 2000 yards, the vacuum tube radios were heavy and bulky. The American army devised a "horse-pack set," which mounted on a horse and resembled a saddle in size and appearance. The radio had its own hand-cranked generator to provide power. The radio still proved a valuable asset, as radio operators could warn of impending gas attacks and troop movements. A radioed warning of a gas attack could give soldiers the time they needed to don protective gas masks. Wired telephone sets were still the most reliable means of battlefield transmission. Some airplanes, themselves a new battlefield development, had radios installed in them. These radios also had limited range and reliability. The radio was most valuable with the navies, as ships at sea could communicate with base stations.

Automobile-Based Telephone 1920

W. W. Macfarlane demonstrated the first recorded mobile telephone on March 21, 1920 in Philadelphia. The second vice president of the Philadelphia based American Dyewood Company held a demonstration for reporters of the Sandusky Register. Mr. Macfarlane sat in the rear seat of his car with his chauffeur driving. He called his wife, who was seated about 500 yards away. He then talked to her on the walkie-talkie type device. He noted that if soldiers in the field during the recently concluded First World War had had the device, it might have saved thousands of lives. The technology would find use twenty years later during World War II when portable two-way radios played a major role in military operations.

1924 - First Two-Way, Voice-Based Radio System

Bell Laboratories tested the first successful two-way, voice based radio system. The equipment was bulky and expensive, but it paved the way for emergency radio systems and the modern cell phone.

1927 - Federal Radio Commission

Congress passed the Radio Act of 1927, which superseded the Radio Act of 1912. The Act established the Federal Radio Commission, which had the power to regulate radio transmissions" as the public interest, convenience, or necessity," required. The Federal Radio Commission existed until 1934, when Congress enacted the Communications Act, which established the Federal Communications Commission.

Donald Lewis Hings Invents the Walkie Talkie

Donald Lewis Hings invented the device called the walkie-talkie in 1937.

Donald Lewis Hings (November 6, 1907 - February 25, 2004)

Native to Leicester, England, Donald migrated to Canada when he was three years old. In 1925, he moved to New Westminster in British Columbia, where he enrolled in a two-year wireless radio course. After completing the course, he worked at a plywood factory. In 1931, the Consolidated Mining and Smelting Company of Canada Limited employed him. The company needed a wireless radio system to communicate with its workers. Hings developed a portable two-way radio by 1937. He called this radio a "packset," a name later changed to walkie-talkie. He filed his patent in 1939. The Canadian government declared war on Germany while he was filing his patent. The Consolidated Mining and Smelting Company of Canada sent him to Ottawa to improve the device for military use. During the war, over 18,000 of the walkie-talkies were used, allowing soldiers on the battlefield to communicate with each other over long distances.

December 11, 1947 - Douglas H. Ring Publishes the First Description of a Cellular Phone Network

Bell Laboratories Douglas H. Ring publishes the first paper that describes the concept of a cellular phone network that could connect automobile drivers with each other on December 11, 1947. Ring's proposal would not be developed for several years; however, his paper laid the groundwork for it.

Douglas H. Ring (March 28, 1907 - September 8, 2000)

Native to Montana, Ring was part of a team of Bell Laboratories' engineers that developed the first cell phones.

1947 - Transistor Invented

Electronic radios, televisions and other equipment needed a device to amplify signals and switch current. Up until 1947, the only thing available was the vacuum tube, first invented in 1906. Vacuum tubes used a lot of power, emitted copious quantities of heat, were expensive and were slow to heat up. Engineers and scientists searched for an alternative. A team of Bell Laboratories engineers came up with the transistor. The team included John Bardeen, Walter Brattain, and William Shockley.

The Transistor

Much smaller than a vacuum tube, the transistor consumes much less power, and does the job a tube does more efficiently. The transistor is still the main component of modern electronics devices. The transistor is usually made from pure silicone or germanium, and has three terminals that connect to an outside circuit.

1946 - AT&T's Mobile Telephone Service

AT&T had been using a mobile telephone service to its own workers before World War II. In 1946, they began offering it to the public. The first mobile telephones were nothing like the hand held units in use currently. The equipment weighed about eighty pounds and consisted of an antenna, battery, battery, receiver, hand held phone unit and control cable. To make a call, the caller had to connect with an operator to make the connection and hold a button down to talk and released to hear the reply. By 1949, the system had about 5,000 subscribers. The service handled about 117,000 calls per year in about sixty cities in the United States and Canada.

Radio Common Carrier

Independent telephone companies attempted to compete with AT&T's Mobile Telephone Service in the early 1960's with a system they called the Radio Common Carrier. Similar to a walkie-talkie, the phones in the system communicated with each other using radio waves. The caller had to depress a button to talk and release the button to listen. These phones did have their own telephone numbers. This was a loosely organized system. Most phones would not work outside their area, thus could not "roam" like modern cell phones. The later phones using this system could roam to a limited degree, but this was not encouraged, as there was no centralized billing system, making billing for calls complicated. The equipment was bulky and expensive. This system persisted until the 1980's, when the first true cellular networks appeared.

1964 New Improved Mobile Telephone Service

In 1964, mobile phone providers began offering the Improved Mobile Telephone Service (IMTS). This system connected to the public switched telephone network and offered direct dial instead of having to dial an operator. The system gradually replaced the Mobile Telephone Service introduced in 1946. These pre cellular phones required a nineteen-inch long antenna to be installed on the car. A heavy cable running from the battery supplied the power needs of this energy hungry device. If used extensively without the motor running, the phone system would run the battery down. The units were expensive, costing $2,000 to $4,000 each. The technology required the providers to limit the number of subscribers; consequently, many potential subscribers were placed on waiting lists and could wait years before receiving their subscription.

April 3, 1973 - Martin Cooper Demonstrates Hand Held Telephone in Public First Time

Dr. Martin Cooper of Motorola performed the first public demonstration of a hand held mobile telephon on April 3, 1973.

Martin Cooper (born December 26, 1928)

The son of Mary Cooper and Arthur Cooper, Martin was native to Chicago, Illinois. He received his elementary education at Lawson Elementary School in Chicago, and then attended Illinois Institute of Technology. He graduated in 1950 with a Bachelor of Science in Electrical Engineering. To finance his education, Cooper had enlisted in the Reserve Officer Training Corps and after graduation; he fulfilled his military obligation by serving on a United States naval destroyer. While in the Navy, he attended submarine school, after which he served on the submarine U.S.S. Tang. After completing his military service, Cooper joined the Teletype Corporation, a subsidiary of American Telephone & Telegraph (AT&T). Cooper worked for Teletype until 1954, when he left to work for Motorola. Cooper became involved with Motorola's development group for mobile devices, a career track that led to the development of the first hand held device that he demonstrated on April 3, 1973.

The Phone

The telephone measured 9 inches long, 5 inches deep and 1.75 quarters inches wide. The device weighed 2.4 pounds. The battery took 10 hours to charge and provided about thirty minutes of talk time. From this telephone, dubbed Zero Generation, or 0G, the hand held device has evolved into the modern cell phone we use today.

1983 - The First Cellular Network

Ameritech Mobile Communications established the first cellular network of mobile telephone service in Chicago on April 13, 1983. A geographic region served by one cell phone tower is called a "cell." A cellular network exists when subscribers using one cell tower can call phones connected to users on other cell towers. A cell tower can both receive and send radio signals, thus allowing freedom of communication between subscribers. Each cell operates on a slightly different radio frequency and allows callers to travel between cells with little or no service interruption.

1G Cellular Networks

1G Cellular Networks are known as the First Generation of cellular phone networks. Introduced in the 1980's, the 1G network was an analog radio signal. The analog system existed, with upgrades, until 2G networks replaced them, beginning around 1991.

Motorola DynaTAC

Motorola released the first commercially available handheld mobile phone, the Dynamic Adaptive Total Area Coverage (DynaTAC) telephone in 1983. The phone cost $3,995, took ten hours to charge and allowed users to talk about thirty minutes to a charge. The phone weighed almost two pounds, measured thirteen inches tall. It could store thirty numbers in its memory. Motorola continued to offer the DynaTAC, with upgrades, until the MicroTAC replaced it, beginning in 1989.

2G Cellular Networks

Initially launched in 1991 in Finland, 2G Cellular Networks featured digital signaling instead of analog, like the 1G. Digital signaling is more efficient than analog. The new technology allowed new services like text messaging and photo and multimedia transmission. 3G superseded 2G, beginning in 1998.

Motorola MicroTAC

Motorola released the MicroTAC 9800X on April 25, 1989. The first of the so called "flip phones," the MicroTAC was diminutive compared to its forerunner, the DynaTAC. TheMicroTAC was only nine inches long when flipped open and would easily fit into a shirt pocket. The phone weighed just over twelve ounces. The phone featured a red dot-matrix LED display and retailed for about $2500.00. Motorola continued production of this phone, with upgrades, until 1998. Due to software problems and battery life, the phone only remained available for six months. IBM sold about 50,000 Simon phones during its time on the market.

1992 First Smart Phone Invented

IBM invented the first cellular phone to be called a "smart phone," in 1992; however, people did not start referring to the new phones by that name for three more years. Nicknamed "Simon," allowed users to send telephone calls, facsimiles, emails and cellular pages. The phone also featured a calendar, address book and notepad. It could also show users maps, stock quotes and news stories.

3G Cellular Networks

Introduced in 1998, 3G Cellular Networks offered at least peak data rates of at least 200 kbit/s (about 0.2 Mbit/s). The improved networks provide wireless voice telephony, mobile Internet access, fixed wireless Internet access, video calls and mobile TV.

2000 - First Camera Phone Released

Samsung released the first camera phone, called the SCH-V200, in South Korea in 2000. This device was essentially a separate phone and telephone within one body. After taking photos, the user had to hook the camera to a computer to download the photos. The device only held 20 .35 mega pixel photos. Sharp introduced a camera phone, the J-SH04, in November the same year that began to approach the modern camera phone. The phone allowed users to send the photos electronically. Sanyo introduced the SCP-5300 in the United States in 2002. The phone had a digital timer, took 640 X 480 mega pixel photos, a flash, a zoom and various photo filters. The phone was heftier than modern ones, but displayed many of the features of today's models.

2007 – Steve Jobs Introduces the Apple I-phone

Apple founder Steve Jobs and Steve Wozniak announced the release of the I Phone on June 29, 2007. The new phone sported several new features, such as internet access, and quickly became popular. The phone had two models, a 4 gigabyte and an 8-gigabyte model priced at $499 and $500 respectively. Market analysts estimate that between 250,000 and 700,000 I Phones sold during the first weekend they appeared and that 146,000 customers activated their phones during that time. Apple has continued to upgrade the I Phone and it remains one of the most popular cellular phones among consumers.

Steve Jobs (February 24, 1955 – October 5, 2011)

The natural son of Joanne Schieble (later Joanne Simpson) and Abdulfattah "John" Jandali, Steven Jobs was native to San Francisco, California.

Education

The two were students of the University of Wisconsin and soon gave the baby boy up for adoption. Paul and Clara Jobs adopted the boy in 1957 and named him Steven Paul Jobs. The couple moved to Mountain View, California where Steven attended Monta Loma Elementary. His father owned an electronics store where Steve learned how to work on electronics. A bright, but poor student, the routine of school bored the boy. He later attended Homestead High School in Sunnyvale, California. After graduation, Jobs enrolled in Reed College in Portland, Oregon.

Early Career

He worked at various jobs, including a stint at Atari, a company that designed video games. He left that company in 1974 to go on a spiritual quest to India. He returned to the United States and became reacquainted with a friend Jobs had met while he was in high school, Steve Wozniak. Wozniak had designed a new computer. Jobs liked the design and suggested that he and Wozniak market the computer and sell it. Wozniak agree, so the two established Apple Computer in Job's garage.

Apple Computer

The name derives from some time Jobs spent in a farm commune orchard in Oregon. Jobs sold his Volkswagen bus and Wozniak a prized scientific calculator to finance the new company. Wozniac's computer revolutionized the computer industry because it was smaller, cheaper, intuitive and accessible to everyday consumers that computers on the market at the time. The computer was a success, earning the two men a fortune over the years.

Jobs Departs Apple, then Returns

Jobs left Apple in 1986 to found Pixar Animation Studios, a company that he had purchased from George Lucas. This company produced many hit movies over the next several years. Meanwhile, Apple had slumped, so Jobs returned in 1997 to help revitalize the company he had helped found. Their new products, the Macbook Air, I Pod and I Phone helped reestablish the company as a leader in its industry. Jobs died of cancer in 2011.

Steve Wozniak (August 11, 1950)

The son of the son of Francis Jacob "Jerry" and Margaret Louise Kern Wozniak, Steve was native to San Jose, California.

Education

During his childhood, his favorite television show was Star Trek. The computer technology depicted in the show inspired his later development of the Apple Computer. Wozniak attended Homestead High School and went on to enroll at University of Colorado. The University expelled him for misuse of the computer system, so Wozniak returned to California to attend De Anza College and later the University of Berkeley.

First Computer

Wozniak, though not a brilliant student, had an uncanny knack for building electronic devises from scratch. Using this skill, he built his first computer that he called the "Cream Soda Computer," in 1971. A mutual friend suggested that he meet another young man, Steven Jobs, who also liked to work on electronics. Jobs, still in high school, met Wozniak and their friendship would launch Apple Computer. Wozniak began working on his new computer while Jobs was in India.

Apple

When Jobs came back, he suggested that the two form a company with Wozniak building the computer and components, while Jobs handled building the case and marketing their product. The product's success led to Wozniak designing the Apple II.

Departing Apple

Wozniak worked for Apple until 1985, when he returned to University of California Berkeley to complete his degree. Wozniak spent time teaching and doing various philanthropic and business ventures. President Ronald Reagan awarded Jobs and Wozniak the National Medal of Technology. Wozniak has received many other awards and honorary degrees including induction into the National Inventors Hall of Fame in 2000.

4G Cellular Networks

Sprint began offering 4G technology in 2009. 4G service includes mobile web access, IP telephony, gaming services, high-definition mobile TV, video conferencing, and 3D television. To date, June 2018, 5G technology has not been introduced.

Satellite Mobile Phones

Satellite phones offer the advantage of operating in places not served by a cellular network. These telephones can make phones calls from extremely remote locations because they connect directly to satellites orbiting 22,500 miles above the earth.

The Age of Satellite Communications

The age of satellite communications opened when the National Aeronautics and Space Administration (NASA) launched the first telecommunications satellite, Telstar I, using a Thor-Delta rocket on July 10, 1962. President Lyndon Johnson made the first official satellite phone call later that day. Telstar also transmitted the first television images and a satellite fax transmission.

Telstar I

Built by American Telegraph and Telephone Company (AT&T), the satellite's name derives from a combination of two words, telephone and star. Weighing in at about 170 pounds, Telstar I had an elliptical orbiting at an altitude of 592 miles at perigee and 3,687 at apogee. The three-foot long satellite orbited about every two hours and forty minutes, providing only about twenty minutes of service before moving out of communication. The satellite used 3600 solar panels on the hull to generate about fourteen watts of power. The satellite had a capacity of about 600 phone calls and one television signal. The low radio strength required huge receiving stations on the ground to pick up the signal. Telstar I did not last long. An earlier atmospheric nuclear weapons test conducted by the United States had energized the Van Allen belt that surrounds the earth. The radiation from this test, as well as a later Soviet Russian test, fried the circuits of Telstar, forcing it out of commission in November 1962. Telstar fluttered into life briefly in February 1963, then went dead. Telstar II launched in 1964. Today there are over 500 active satellites in earth orbit with thousands more that have gone inactive. The Telstar satellites still orbit the earth, though they have long been dead. Today's communications satellites maintain a geosynchronous orbit of about 22,500 miles above the earth, meaning that their orbital speed is the same as earth's rotational speed at that altitude and the satellites remain on one spot over the earth allowing it to remain in continuous communication.

Fiber Optic Cables

Fiber optic cables consist of fine strands of silica (glass) drawn to approximately the thickness of a human hair. The fibers transmit light that contains data between two points. Fiber optics has replaced copper cables in many applications because they transmit data faster and with less loss over distances. Fiber optics is also less prone to electromagnetic interference than copper cables are. The fiber optics can bend light as it courses along its length. This property, known as refraction, was first demonstrated in the early Nineteenth Century. Advances in the technology opened its use as a means of exploring inside the human body to illuminate areas for medical examination. Further advances in the 1960's and 1970's led to its use as a means of transmitting telephone and data signals by the 1980's. The installation of fiber optics replacing copper cables has continued to improve communication speed and efficiency.

Powering the Telephone

History of the Battery

During the middle part of the Eighteenth Century, scientists began to explore the properties of the mysterious force called electricity. The ancient Greeks knew this force. They discovered that by rubbing two pieces of amber together they could make static electricity. Indeed, the name electricity comes from the Greek work for amber, elektron. Modern devices require compact, powerful batteries, thus the history of the battery becomes part of the history of modern mobile technology.

Leyden Jars

The first devices used to store electricity were called Leyden Jars. German cleric Ewald Georg von Kleist and Dutch scientist Pieter van Musschenbroek each independently invented this device. Kleist built his on October 11,1745. Musschenbroek developed his during the same period. Musschenbroek lived in Leiden (Leyden), South Holland and performed his electricity experiments there. The Leyden jar is not a true battery. It stores static electricity, looses its charge over a short time and does not discharge consistently. It is really more like an electrical device called a capacitor. Musschenbroek publicized his experiments with the Leyden jar to the French scientific community. The French scientists subsequently referred to the new device as a "Leyden Jar." Ben Franklin performed much important work in electricity and used Leyden Jars to store electricity for his experiments. He hooked several of the jars together in a series and managed to produce an increased amount of electricity from them. He coined the term "battery" comparing it to a battery of cannon.

The First Battery

Alessandro Volta, an Italian physicist and chemist, developed the first true battery in 1800. Known as a Voltaic pile, the device consisted of several copper and zinc paired discs. A brine soaked piece of cardboard or cloth separated the discs. The device produced an electric current when a wire connected the top and bottom contacts. His battery became known as a wet pile, because of the reliance of liquid brine to conduct the electricity. Volta later experimented with wet pile devices whose cardboard separators had dried out. The device worked, and came to be known as a dry pile. The modern battery evolved out of the dry pile battery.

Trough Battery

Two problems associated with Volta's battery were electrolyte leaks and short battery life. Scotsman William Cruickshank solved this by placing the coils inside a box instead of piling them vertically. Hydrogen bubbles forming in the copper caused short battery life. This increased electrical resistance and reduced the life of the battery.

Daniell cell

British chemist John Frederic Daniell used a copper pot filled with a copper sulfate solution. This pot was immersed an unglazed earthenware container. The container contained with sulfuric acid and a zinc electrode. This Daniell cell, as it came to be known, solved the hydrogen problem and produced a battery with increased life. He developed this device in 1836.

Telephone Invention Spurs Battery Technology

Most inventors and scientists considered the battery as little more than a toy until the telegraph, and later the telephone, appeared. Both devices needed a local source of electrical power that only a battery could supply. The first telephones each had their own battery. One of the jobs for early employees of a telephone company was to visit each subscriber to check the condition of the battery.

The Modern Battery Emerges

Over the next several decades, other scientists improved the battery, each adding their own knowledge to it. By In 1887, Yai Sakizo, a Japanese inventor, developed a dry battery. The first rechargeable nickel-cadmium rechargeable followed in 1899, developed by Swedish scientist named Waldemar Jungner. Thus, by 1900 the foundation for the modern battery had evolved from the original Leyden Jars in 1745.

Local Battery Power to Common Battery Power

Local battery power systems required each telephone to have its own battery. By the mid-1890's some telephone companies developed, a new system called the Common Battery Power system. In these systems, the battery that powered the phone was located in the telephone exchange with power transmitted to the telephone over a copper wire. Since the common battery system replaced the local battery on each telephone, designers could now design smaller, more compact telephones. The current supplied to the telephones was a low voltage current. Telephone companies still supply the electric to telephones using large, rechargeable battery banks, which is why telephone systems continue to operate if there is a power outage.

Nickel Cadmium (Ni-Cd)

Swedish Inventor Waldemar Jungner developed the Ni-Cd in 1899. The Ni-Cd did not see much use in the United States until 1946. By the mid 1990's, Ni-Cds dominated the rechargeable battery market in the United States. The batteries are available in most of the popular consumer battery sizes. Around 1992 both Lithium Ion and Nickel Metal Hydride became economical and the use of Ni-Cd's declined. The Ni-Cd battery saw wide use in electronics, tools and consumer batteries in the 1970's and 1980's. They are more economical than disposable cells, but they do have some disadvantages. These cells can develop a "memory." If recharged after only a partial discharge several times, they will not discharge all the power stored in them. They also have a high discharge rate of ten percent a month if unused.

Charging

Charging time for Ni-Cds depends upon the type of battery. Some will take "fast chargers" of one hour or less and others will not. It is best to read the battery's charging requirements before putting it in a charger. Many of the electronic devices that still rely on Ni-Cds will probably supply a special charger for the included battery. Follow manufacturer's recommendations for these batteries. Battery chargers will supply power to the battery at a slightly higher rate than the battery will contain when fully charged for a rate of time needed to charge the battery. This time can vary according to the charger and battery type. Make sure you use a charger compatible with Ni-Cds. Use of an incompatible charger can damage the battery or cause it to explode. It is normal for the battery to increase in temperature while charging. Many automatic chargers use this temperature increase as a signal to stop charging. When the battery reaches a certain temperature, it is charged and the charger ceases. Leaving batteries in a charger after charging is complete can damage the battery or reduce its life. Typically, a Ni-Cd battery can take hundreds of recharges during its lifetime, as long as manufacturer's recommendations are followed.

Materials Used and Disposal

The electrodes consist of nickel oxide hydroxide and metallic cadmium as electrodes. The name Ni-Cd derives from the Periodic Table symbols of nickel (Ni) and cadmium (cd). The Saft Groupe S.A. Company has registered the name NiCad as a copyrighted trademark. The heavy metals involved in their manufacture also became a matter of concern. The disposal of large numbers of them could harm the environment. The Ni-Cd is 100% recyclable. The Rechargeable Battery Recycling Corporation (RBRC) organizes and promotes rechargeable battery collection in the United States and Canada. Click the link to find a battery collection point near you.

http://www.rbrc.com/

Nickel Metal Hydride Batteries (Ni–MH)

Developed in 1967, the Ni–MH cell came on the market in 1989. By 1990, the Nickel Metal Hydride started stealing market share from Ni-Cd. Presently they have almost completely superseded them. They do not develop a memory, have higher storage capacity and discharge much more slowly when not in use. The materials used to manufacture are also less of an environmental concern. This is especially true if consumers recycle them.

The Ni–MH carries two to three times the electrical capacity of similar sized Ni-Cd batteries. Chargers typically use 1.4 to 1.6 volts to charge a Ni–MH battery. A smart charger should be used if fast charging the battery. Do not use a Ni-Cd charger to charge a Ni–MH battery. A fully charged Ni–MH can outperform a similar sized alkaline battery. The starting voltage of a newly charged Ni–MH battery is about 1.4 volts, versus the 1.2 volts of a Ni-Cd. Using low end chargers may result in an unsatisfactory charge, which leads to shorter life in the re-charged battery. This can create the need for more frequent charging. For best results, remove the batteries from the charger when charging is complete to prevent overcharge, which shortens battery life. Charging requirements can vary due to the battery's manufacturer.

Always follow the manufacturer's recommendations when charging a battery.

Standard Ni–MH batteries have an average self-discharge rate of around one to four percent, lower than a Ni-Cd. This rate can change with room temperature. Newer Ni–MH batteries came on the market in 2005. Called low self-discharge nickel–metal hydride battery (LSD NiMH), these batteries retain 75 - 85% of their charge one year from charging if stored at sixty-eight degrees Fahrenheit. The trade-off with these batteries is that they have a lower capacity than standard Ni-MH cells.

The materials used to manufacture Ni–MH batteries are much more environmentally friendly than Ni-Cds. The toxins used are mild and the batteries are recyclable.

Lithium Ion (Li-Ion)

The research and development of these batteries occurred during the 1980's after a researcher first proposed them in 1973. The first ones entered the marketplace in 1991 when Sony and Asahi Kasei introduced the first ones. They find use in cell phones, computers and other electronic devices that need a compact, powerful battery. The battery in your I Pad or cell phone is most likely a Lithium Ion.

Construction of the Battery

The three functioning parts of a lithium Ion battery are positive and negative electrodes and an electrolyte. Carbon comprises the negative element, a metal oxide the positive one. Lithium salt suspended in an organic solvent is the electrolyte. The lithium salts are combustible under the right conditions. Because of this, and the fact that they are kept under pressure, they can present a safety issue. Lithium Ion batteries have been the subject of accidents and recalls over the years manufacturers have used them.

Uses

Li-Ion batteries offer high energy density, low discharge rate and lightweight as their main advantage over other kinds of batteries. Because of these properties the Li-Ion battery is currently the predominate battery type used in cell phones, computers, electronic games and other electronic devices. They can take quick recharge and have a long battery life for most devices. Manufacturers are starting to replace typical lead-acid batteries with Lithium Ion in golf carts and electric powered utility vehicles.

Charge and Discharge

Most devices that use this type of battery supply the correct charger for the battery in the device. Always follow manufacturer's instructions for charging and do not use a charger meant for another size battery to charge. The charging voltage may not be correct. The charger will apply a slightly higher voltage than the battery delivers. The best charging temperatures for a Lithium Ion battery is between 41 to 113 ° Fahrenheit. Do not charge one of these batteries when it is exposed to 32 ° or less. Most devices will not charge at low temperature. However, cell phones may allow limited charging if they detect an emergency call being made. Most of these batteries self discharge at a rate of about 1.5% - 2% per month when not in use.

Battery Life

Most device batteries have a useful life of around three years. Storage and the total number of charge/discharge cycles will affect this time. If the battery is stored, unused, it may reduce the batteries life as will the total charge/discharge cycles. Most will last for about 500 - 1000 cycles, although some carbon based anode types may last 10,000 cycles.

Acknowledgments

Telephone Etymology

https://www.etymonline.com/word/tele-?ref=etymonline_crossreference

Lover's Phone

https://en.wikipedia.org/wiki/Tin_can_telephone

https://bebusinessed.com/history/history-of-the-telephone/

Robert Hooke

https://en.wikipedia.org/wiki/Robert_Hooke

"Solrésol"

https://en.wikipedia.org/wiki/François_Sudre_(1787–1862)

http://mentalfloss.com/article/77536/original-telephone-bizarre-musical-language-jean-francois-sudre

Jean-François Sudré (August 15, 1787 - October 3, 1862)

https://ipfs.io/ipfs/QmXoypizjW3WknFiJnKLwHCnL72vedxjQkDDP1mXWo6uco/wiki/François_Sudre.html

https://en.wikipedia.org/wiki/François_Sudre_(1787–1862)

http://oxfordindex.oup.com/view/10.1093/gmo/9781561592630.article.46105

Banvard's Folly: Thirteen Tales of Renowned Obscurity, Famous Anonymity, and ...

By Paul Collins

https://books.google.com/books?id=sIGpBgAAQBAJ&pg=PA105&lpg=PA105&dq=Jean-François+Sudré+died+1862&source=bl&ots=gNAsLwCyVY&sig=5-U2xbfWPD2Pp0jzkqiDGgsrRdo&hl=en&sa=X&ved=0ahUKEwj21rjpqcTZAhUV24MKHfxiBwYQ6AEIbjAR#v=onepage&q=Jean-Fran%C3%A7ois%20Sudr%C3%A9%20died%201862&f=false

Semaphore Systems

http://people.seas.harvard.edu/~jones/cscie129/papers/Early_History_of_Data_Networks/Chapter_2.pdf

https://www.britannica.com/technology/semaphore

https://en.wikipedia.org/wiki/Semaphore_line

Hidden Codes & Grand Designs: Secret Languages from Ancient Times to Modern Day

By Pierre Berloquin

https://books.google.com/books?id=F9q8BAsXTWEC&pg=PA26&lpg=PA26&dq=first+optical+line+telegraph+boston+martha%27s+vineyard&source=bl&ots=fD8nNXH6-U&sig=iYCu68ZScbioT4C1On0dhE_Fa8o&hl=en&sa=X&ved=0ahUKEwjI7qeIucbZAhWS

wYMKHdLiAkAQ6AEITDAE#v=onepage&q=first%20optical%20line%20telegraph%20boston%20martha's%20vineyard&f=false

Early Telegraph Systems

http://people.seas.harvard.edu/~jones/cscie129/images/history/lesage.html

http://www.rochelleforrester.ac.nz/electric-telegraph.html

https://en.wikisource.org/wiki/Morrison,_Charles_(DNB00)

The First Telegraph

An Historical Sketch of Henry's Contribution to the Electro-magnetic

By William Bower Taylor

https://books.google.com/books?id=D-YOAAAAYAAJ&pg=PA9&lpg=PA9&dq=salva+telegraph+madrid+1798&source=bl&ots=ONHIAFNa8H&sig=vuxg_8-ZUWsipkP8WSiSS1HzCiA&hl=en&sa=X&ved=0ahUKEwi58tnP-9fZAhVF7IMKHduUBPkQ6AEIKTAA#v=onepage&q=salva%20telegraph%20madrid%201798&f=false

https://en.wikipedia.org/wiki/Voltaic_pile

https://www.semanticscholar.org/paper/Early-proposals-of-wireless-telegraphy-in-Spain%3A-F-Romeu-Elias/f27c41949273e7494f2e0ff2617ad7cce7fe5329

https://en.wikipedia.org/wiki/Francisco_Salva_Campillo

http://www.dmg-lib.org/dmglib/main/biogrViewer_content.jsp?id=9003004&skipSearchBar=1

https://thebiography.us/en/salva-y-campillo-francisco

Morse's Electrical Telegraph

http://reference.insulators.info/publications/view/?id=5436

https://en.wikipedia.org/wiki/Samuel_Morse

http://www.in.gov/library/files/S421_Ellsworth_Annie_Goodrich_Collection.pdf

https://www.loc.gov/collections/samuel-morse-papers/articles-and-essays/invention-of-the-telegraph/

Samuel von Sommering's Telegraph

http://users.manchester.edu/FacStaff/SSNaragon/Kant/bio/FullBio/SoemmerringST.html

http://ethw.org/Samuel_Thomas_von_Sömmerring

https://en.wikipedia.org/wiki/Samuel_Thomas_von_Sömmerring

Innocenzo Manzetti's "Speaking Telegraph"

http://www.manzetti.eu/home/the-real-telephone-inventor/

https://en.wikipedia.org/wiki/Innocenzo_Manzetti

http://www.isemag.com/2015/03/innocenzo-manzetti/

http://bizzarrobazar.com/en/2015/07/15/speciale-innocenzo-manzetti/

Johann Philipp Reis (January 7, 1834 - January 14, 1874)

http://www.telephonecollecting.org/Bobs%20phones/Pages/Essays/Reis/Reis.htm

https://en.wikipedia.org/wiki/Johann_Philipp_Reis

https://www.washingtonpost.com/lifestyle/kidspost/first-telephone-by-johann-philipp-reis-was-laughed-off-as-a-toy/2011/10/31/gIQAm4TZ2N_story.html?utm_term=.c6b8335e836e

https://en.wikipedia.org/wiki/Reis_telephone

Antonio Meucc The First Telephone?

https://en.wikipedia.org/wiki/Antonio_Meucci

https://www.famousscientists.org/antonio-meucci/

http://www.italianhistorical.org/page42.html

Alexander Graham Bell and the First Telephone

https://en.wikipedia.org/wiki/Alexander_Graham_Bell

http://www.pbs.org/transistor/album1/addlbios/bellag.html

https://www.famousscientists.org/alexander-graham-bell/

http://www.antiquetelephonehistory.com/telworks.php

https://en.wikipedia.org/wiki/Invention_of_the_telephone

1877 - First Long-Distance Telephone Line

https://en.wikipedia.org/wiki/First_long-distance_telephone_line

https://www.sierranevadageotourism.org/content/worlds-first-long-distance-telephone-line-no-247-california-historical-landmark/sie0ea694a71767f084f

Elisha Gray's Telephone

https://en.wikipedia.org/wiki/Elisha_Gray

https://www.britannica.com/biography/Elisha-Gray

http://www.ohiohistorycentral.org/w/Elisha_Gray

https://www.geni.com/people/Elisha-Gray/6000000013078998166

Water Microphone

https://en.wikipedia.org/wiki/Water_microphone

https://en.wikipedia.org/wiki/Thomas_Edison

Pulsion Telephone

https://paperspast.natlib.govt.nz/newspapers/HBH18900130.2.17

https://scripophily.net/imputesecoma.html

Carbon Microphone

http://www.streetdirectory.com/travel_guide/114823/phones/the_basic_information_on_the_carbon_microphone.html

https://en.wikipedia.org/wiki/Thomas_Edison

https://www.biography.com/people/thomas-edison-9284349

Sound-Powered Telephones

https://www.nps.gov/nhl/find/withdrawn/telephone.htm

https://en.wikipedia.org/wiki/History_of_the_telephone

Early Ringers

https://www.telecom-milestones.com/telephone-history

Strowger Switch

https://en.wikipedia.org/wiki/Strowger_switch

http://www.slate.com/blogs/the_eye/2013/10/03/strowger_switch_the_19th_century_design_invention_that_flipped_the_phone.html

Research Paper by Bob Stoffels www.snohomishhistoricalsociety.org/wp/wp-content/uploads/2017/06/strowger.doc

Rotary Dial

https://www.telecom-milestones.com/telephone-history

Candlestick Telephone

https://en.wikipedia.org/wiki/Candlestick_telephone

First Transcontinental Telephone Call

https://en.wikipedia.org/wiki/History_of_the_telephone

https://www.nationalgeographic.org/thisday/jan25/first-transcontinental-telephone-call/

Radiotelephone

https://en.wikipedia.org/wiki/Radiotelephone

Bell's Model 102 Telephone

https://ipfs.io/ipfs/QmXoypizjW3WknFiJnKLwHCnL72vedxjQkDDP1mXWo6uco/wiki/Model_102_telephone.html

Push Button Telephone

https://en.wikipedia.org/wiki/Push-button_telephone

Party Lines

https://en.wikipedia.org/wiki/Party_line_(telephony)

Cellular Phone History

https://bebusinessed.com/history/history-cell-phones/

https://en.wikipedia.org/wiki/History_of_mobile_phones

https://www.soneticscorp.com/history-of-two-way-wireless/

Electromagnetic Wave Transmission

http://www.seas.columbia.edu/marconi/history.html

https://eandt.theiet.org/content/articles/2014/06/ww1-first-world-war-communications-and-the-tele-net-of-things/

https://www.uswitch.com/mobiles/guides/history-of-mobile-phones/

https://en.wikipedia.org/wiki/Nikola_Tesla

https://www.nobelprize.org/nobel_prizes/physics/laureates/1909/marconi-bio.html

https://en.wikipedia.org/wiki/Electromagnetic_radiation

https://www.soneticscorp.com/history-of-two-way-wireless/

https://www.geni.com/people/Heinrich-Rudolf-Hertz/6000000015888101035

https://www.famousscientists.org/heinrich-hertz/

https://en.wikipedia.org/wiki/Heinrich_Hertz

http://thepandorasociety.com/whats-in-the-box/

Audion Vacuum Tube

https://en.wikipedia.org/wiki/Control_grid

https://www.electronics-notes.com/articles/electronic_components/valves-tubes/electrodes-cathode-grid-anode.php

http://www.leedeforest.org/The_Audion.html

http://www.vacuumtubes.net/How_Vacuum_Tubes_Work.htm

Lee DeForest (August 26, 1873 – June 30, 1961)

https://en.wikipedia.org/wiki/Lee_de_Forest

Radio Act of 1912

https://en.wikipedia.org/wiki/Radio_Act_of_1912

https://www.soneticscorp.com/history-of-two-way-wireless/

https://earlyradiohistory.us/1912act.htm

Radio Act of 1912

Radio in the First World War

https://dp.la/exhibitions/radio-golden-age/radio-frontlines

https://www.livescience.com/45641-science-of-world-war-i-communications.html

Automobile-Based Telephone 1920

https://www.smithsonianmag.com/history/the-worlds-first-carphone-24664499/

Smart Cities: Big Data, Civic Hackers, and the Quest for a New Utopia

https://books.google.com/books?id=dc96AAAAQBAJ&pg=PT41&lpg=PT41&dq=W.+W.+Macfarlane+1920&source=bl&ots=4iPjuI9RKo&sig=VItS3wAJtq_mrvMMnyU7g5TX4F4&hl=en&sa=X&ved=0ahUKEwjo4oHujPbaAhXq5oMKHXtjCqkQ6AEIPTAD#v=onepage&q=W.%20W.%20Macfarlane%201920&f=false

1924 - First Two-Way, Voice-Based Radio System

https://www.soneticscorp.com/history-of-two-way-wireless/

http://tapsbus.com/ip-radios/

1927 - Federal Radio Commission

https://en.wikipedia.org/wiki/Federal_Radio_Commission

Donald Lewis Hings Invents the Walkie Talkie

http://walkietalkieworld.com/how-was-the-walkie-talkie-invented/

https://en.wikipedia.org/wiki/Donald_Hings

http://www.canadianminingjournal.com/features/tales-of-teck-and-cominco/

https://prezi.com/yfnu-qg854te/entrepreneur-donald-hings/

December 11, 1947 - Douglas H. Ring Publishes the First Description of a Cellular Phone Network

https://en.wikipedia.org/wiki/Douglas_H._Ring

https://www.theatlantic.com/technology/archive/2011/09/the-1947-paper-that-first-described-a-cell-phone-network/245222/

1947 - Transistor Invented

https://en.wikipedia.org/wiki/Transistor

http://www.physlink.com/education/askexperts/ae414.cfm

1946 - AT&T's Mobile Telephone Service

http://www.wb6nvh.com/MTSfiles/Carphone1.htm

http://gadgetlabrepair.com/mobile-phone-history/

http://www.wb6nvh.com/MTSfiles/Carphone1.htm

Radio Common Carrier

https://www.popsugar.com/tech/photo-gallery/29019196/image/29019200/Radio-Common-Carrier-RCC

https://en.wikipedia.org/wiki/Mobile_radio_telephone

https://bebusinessed.com/history/history-cell-phones/

1964 New Improved Mobile Telephone Service

https://en.wikipedia.org/wiki/Improved_Mobile_Telephone_Service 1964

1983 - The First Cellular Network

https://www.nationalgeographic.org/thisday/oct13/first-american-cellular-network/

http://www.knowyourmobile.com/nokia/nokia-3310/19848/history-mobile-phones-1973-2008-handsets-made-it-all-happen

Motorola DynaTAC

http://content.time.com/time/specials/packages/article/0,28804,2023689_2023708_2023656,00.html

https://en.wikipedia.org/wiki/Motorola_DynaTAC

Motorola MicroTAC

https://en.wikipedia.org/wiki/Motorola_MicroTAC

http://malibuflash.com/technology/motorola-microtac-9800x

2007 - The Apple Iphone is released

https://en.wikipedia.org/wiki/Steve_Wozniak

https://en.wikipedia.org/wiki/Steve_Jobs

https://en.wikipedia.org/wiki/History_of_iPhone

https://www.biography.com/people/steve-wozniak-9537334

https://www.britannica.com/biography/Stephen-Gary-Wozniak

Satellite Mobile Phones

https://www.jpost.com/Features/In-Thespotlight/This-Week-in-

Fibre Optic Cables

https://en.wikipedia.org/wiki/Optical_fiber

Powering the Telephone

http://ethw.org/Batteries

http://www.moah.org/talkingwires/talkingwires.html?KeepThis=true

https://en.wikipedia.org/wiki/History_of_the_telephone

https://en.wikipedia.org/wiki/History_of_the_battery

http://ethw.org/Telephones

http://batteryuniversity.com/learn/article/charging_nickel_metal_hydride

https://en.wikipedia.org/wiki/Nickel%E2%80%93metal_hydride_battery

https://en.wikipedia.org/wiki/Nickel%E2%80%93cadmium_battery

https://en.wikipedia.org/wiki/Lithium-ion_battery

http://batteryuniversity.com/learn/article/is_lithium_ion_the_ideal_battery

Lithium Ion (Li-Ion)

About the Author

Paul considers himself a bit of an Indiana hound, in that he likes to sniff out the interesting places and history of Indiana and use his books to tell people about them.

Join Paul on Facebook
https://www.facebook.com/Mossy-Feet-Books-474924602565571/
Twitter
https://twitter.com/MossyFeetBooks
mossyfeetbooks@gmail.com

Mossy Feet Books Catalog

To Get Your Free Copy of the Mossy Feet Books Catalogue, Click This Link.

http://mossyfeetbooks.blogspot.com/

Gardening Books

Fantasy Books

Humor

Science Fiction

Semi - Autobiographical Books

Travel Books

Sample Chapter

Timeline of United States History

June 15, 1215 - King John I signs Magna Carta at Runnymede England

For the first time in English history, a band of rebel barons forced a monarch to cede a portion of his power when they compelled King John I to sign the Magna Carta on June 15, 1215. The Magna Carta inspired much of the revolutionary fervor for liberties that resulted in the American Revolution and the Constitution that followed.

King John I Angers the Barons

In 1209, Pope Innocent III had excommunicated King John I over a dispute over the appointment of a new Archbishop of Canterbury. During this period, he also indulged in a disastrous war against the French. His fiscal policies resulted in excessive taxation exacted from the English barons.

Rebellion

English barons had rebelled before; however, they always had a successor in mind. By a matter of coincidences, there was no clear successor available in 1215. The barons had wanted to overthrow John II, but with no successor available, they settled on negotiating with the king. With an armed rebellion brewing, King John II agreed to meet in a meadow near London called Runnymede. At this meeting, the barons presented their grievances to the king. The King signed the document, called the Magna Carta, temporarily resolving the crises.

Continued Defiance

One segment of the Magna Carta (Great Charter) called for the creation of a committee of barons to oversee the King's observance of the Charter. They would have the power to depose the King if they felt he had broken the agreement.

King John II objected to this clause and rebelled against it, prompting the First Baron's War that did not end until King John died in 1216 during a siege. When the new king took power at adulthood, Henry III signed a shorter version of the Charter that did not contain the contentious article.

Influencing Political Thought

The Magna Carta did not confer rights on the common people. It only gave rights to the barons. However, over time the freedoms granted by the Charter expanded and became more important. The men that wrote the Constitution and the Bill of Rights in the aftermath of the American Revolution drew much of their inspiration from the Magna Carta and thus a document signed by a beleaguered English king in 1215 influences United States law today.

Made in the USA
Coppell, TX
18 January 2024